WETLANDS

BIOMES

Lynn M. Stone

The Rourke Corporation, Inc.
Vero Beach, Florida 32964

PHOTO CREDITS
All photos © Lynn M. Stone

Library of Congress Cataloging-in-Publication Data
Stone, Lynn M.
    Wetlands / by Lynn M. Stone.
        p.   cm. — (Biomes)
    Includes index.
    Summary: Describes swamps, marshes, and bogs, places
which are almost always wet and soggy and which are habitats
for hundreds of kinds of plants and animals.
    ISBN 0-86593-425-8
    1. Wetland ecology—Juvenile literature. 2. Wetlands—Juvenile
literature. [1. Wetland ecology. 2. Ecology. 3. Wetlands.] I. Title.
II. Series: Stone, Lynn M. Biomes.
QH541.5.M3S77   1996
574.5'26325—dc20   95-46183
                                                            CIP
                                                            AC

Printed in the USA

TABLE OF CONTENTS

WETLANDS

Even on land, there are places where the ground is almost always wet and soggy. These places are **wetlands** (WET landz). For at least part of the year, they are flooded with shallow water — water that is not very deep.

We call these low, wet areas swamps, marshes, and bogs. They are beautiful and important **habitats** (HAB uh tats), or homes, for hundreds of kinds of plants and animals.

Low tide shows the muddy bottom of a South Carolina salt marsh

KINDS OF WETLANDS

You probably live near a wetland. Wetlands with trees are called swamps.

Open wetlands with cattails and other water-loving plants are freshwater marshes. Saltwater marshes of *Spartina* grass are found in lowlands near the sea.

Bogs often have a pond with trees or dense marsh plants around it. A bog used to be a lake. As the lake plants died and filled it up, the lake became smaller and more shallow.

The state of Florida protects the Fahkahatchee Strand swamp, one of North America's priceless wetlands

WETLAND PLANTS

The roots of wetland plants grow in wet soil. These plants like "wet feet."

Some wetland plants, such as waterlilies, live in water. Other plants — cattails, for example — grow in moist ground near still water.

Several kinds of wildflowers live only in wetlands. One is the insect-eating pitcher plant. The pitcher plant traps insects in its pitcherlike stem.

Long-stemmed waterlilies root in muddy soil below the floating blossoms

LIFE IN THE WETLANDS

Wetland animals need plants — large and small — to live. Plants in a wetland change sunlight into food that helps them grow. An insect, bird, or mammal may eat the plants.

A plant-eating animal, in turn, may be eaten by a larger, fiercer animal — a **predator** (PRED uh tor). In that way, food — and energy — pass along from one living thing to another.

An alligator is about to eat a big fish that ate a smaller fish that ate an even smaller fish that ate plants

Too busy to "stop and talk," a swimming beaver in Alaska
drags an alder branch to its dam

A sandhill crane steps through a freshwater marsh in Florida

BIRDS IN THE WETLANDS

Certain birds are made to live in wetlands. Ducks, for example, paddle about easily with webbed feet. Oil keeps their feathers waterproof.

Herons, **ibises** (I bihss es), storks, and spoonbills have long legs for wading in shallow water. They also have long necks and bills to catch fish and other **aquatic** (uh KWAT ihk), or water, animals.

Grebes are diving birds. They find food in wetlands and build floating nests of wetland plants.

14

The great egret and its cousins are made for wading in wetlands and spearing small animals

MAMMALS OF THE WETLANDS

Several kinds of mammals live in and around wetlands. Many of them — such as river otters, minks, and beavers — have thick, waterproof fur and webbed feet.

The beaver can make its own wetland. Beavers gnaw down trees. They use branches packed with mud to build dams on streams. Then the stream becomes a pond. The deeper water of the pond helps protect the beaver from predators.

The mink's long, lean body and webbed toes help make it an excellent swimmer and fisherman

OTHER WETLAND ANIMALS

In addition to feathered and furry animals, small, boneless animals — like insects or shrimps — live in wetlands. Scaly animals like alligators, turtles, and snakes live in wetlands, too.

Wetlands are important animal nurseries. Young animals of many kinds, including fish, begin their lives in the shallow waters of wetlands.

A toad trills for a marsh mate on a spring night

VISITING THE WETLANDS

You can visit some wetlands on a wooden sidewalk, called a boardwalk. Or you can visit a wetland the way a duck does — with wet feet.

Either way, you may see a muskrat or beaver swimming. You might be buzzed by a dragonfly.

You'll hear the croak of frogs, the hum of insects, and a chorus of bird calls — chirps, whistles, croaks, and quacks.

Be careful if you wade. Deep, sucking mud, hidden branches, and deep pools of water are dangers in the wetland.

This western cattail marsh is home for a nesting trumpeter swan, North America's heaviest water bird

PROTECTING THE WETLANDS

Wetlands are important homes for animals. They also filter the water we use and give people places to explore, fish, hunt, and admire.

Still, wetlands are often destroyed so the land can be used for farming or building. Many people and groups are working hard to keep wetlands wet. Ducks Unlimited, for example, pays farmers in the West to keep marshes. Millions of ducks grow up on these little **potholes** (PAHT holz).

Glossary

aquatic (uh KWAT ihk) — of or related to water, such as an *aquatic* bird

habitat (HAB uh tat) — the kind of place where an animal lives, such as the muddy soil of a *marsh*

ibis (I bihss) — a slender, long-legged wading bird with a long, down-curved bill

pothole (PAHT hole) — a small, roundish pond of the North American prairies

predator (PRED uh tor) — an animal that kills other animals for food

wetland (WET land) — a low, wet area covered by shallow water for at least part of the year; a marsh, bog, or swamp

INDEX